U-211 to Squaw Flats S

35 miles, 2 hours for 70-mile

↑ NORTH
M = milepost

Campground

THE NEEDLES

Red and
Cedar Me:
reminder:
coast, wh
the coast
red sands
over milli
and white
evident ir
courtesy C

Cross-bedded Cedar Mesa Sandstone (right foreground) in Squaw Flats Campground. East view to South Sixshooter Peak (distant left). *Photo courtesy Canyonlands National Park.*

NORTH SIXSHOOTER PEAK

SOUTH SIXSHOOTER PEAK

e Sixshooters
e remnant
ires of Wingate
ndstone.

The area behind South Sixshooter Peak was considered for the nation's first high-level nuclear-waste repository. Salt beds 3,000 feet below the surface were targeted as potential host rocks for the radioactive material.

hite colors within the
a Sandstone are
of the ancient Permian
re white sands from
and inland dunes and
rom rivers mingled
ns of years. The red
bands are particularly
The Needles. *Photo
nyonlands National Park.*

Indian Creek

Shay graben is a downdropped block between two faults. Descending into Indian Creek canyon, the normal progression from younger to older rocks is disrupted. The disorder is most evident on the northwestern fault adjacent to the road, where the red Kayenta Formation is downdropped to the same level as the yellow Wingate Sandstone. Normally the younger Kayenta overlies the older Wingate.

NEWSPAPER ROCK STATE PARK

M7
w   w
k

Shay graben

Shay graben
Simplified cross-section

Newspaper Rock exhibits Indian petroglyphs engraved into desert-varnished Wingate Sandstone.

Beginning at Church Rock, this road crosses the sage flats, winds down to cottonwood-lined Indian Creek canyon, passes intriguing Newspaper Rock, enters a wide valley bounded by Jurassic-age sandstone walls and Triassic-age siltstone slopes, and ends at Squaw Flat Campground in The Needles District of Canyonlands National Park. Dominating the landscape upon entering the park is the Cedar Mesa Sandstone - formerly a Permian coastal area of beaches and inland sand dunes, located between the northeastern river networks and western ocean. Views of the Abajo Mountains, the skyline spires of The Needles, petroglyphs, and rock climbers scaling Wingate Sandstone cliffs add diversity to this drive.

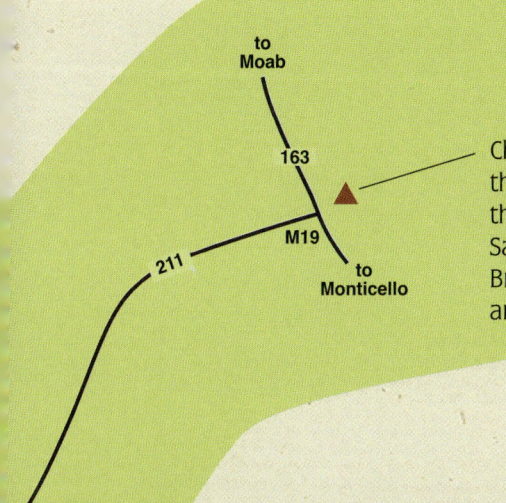

Church Rock stands like a sentinel on the sage flats. The rock displays the three members of the Entrada Sandstone: the basal red Dewey Bridge, the middle yellow Slick Rock, and the upper light-tan Moab Tongue.

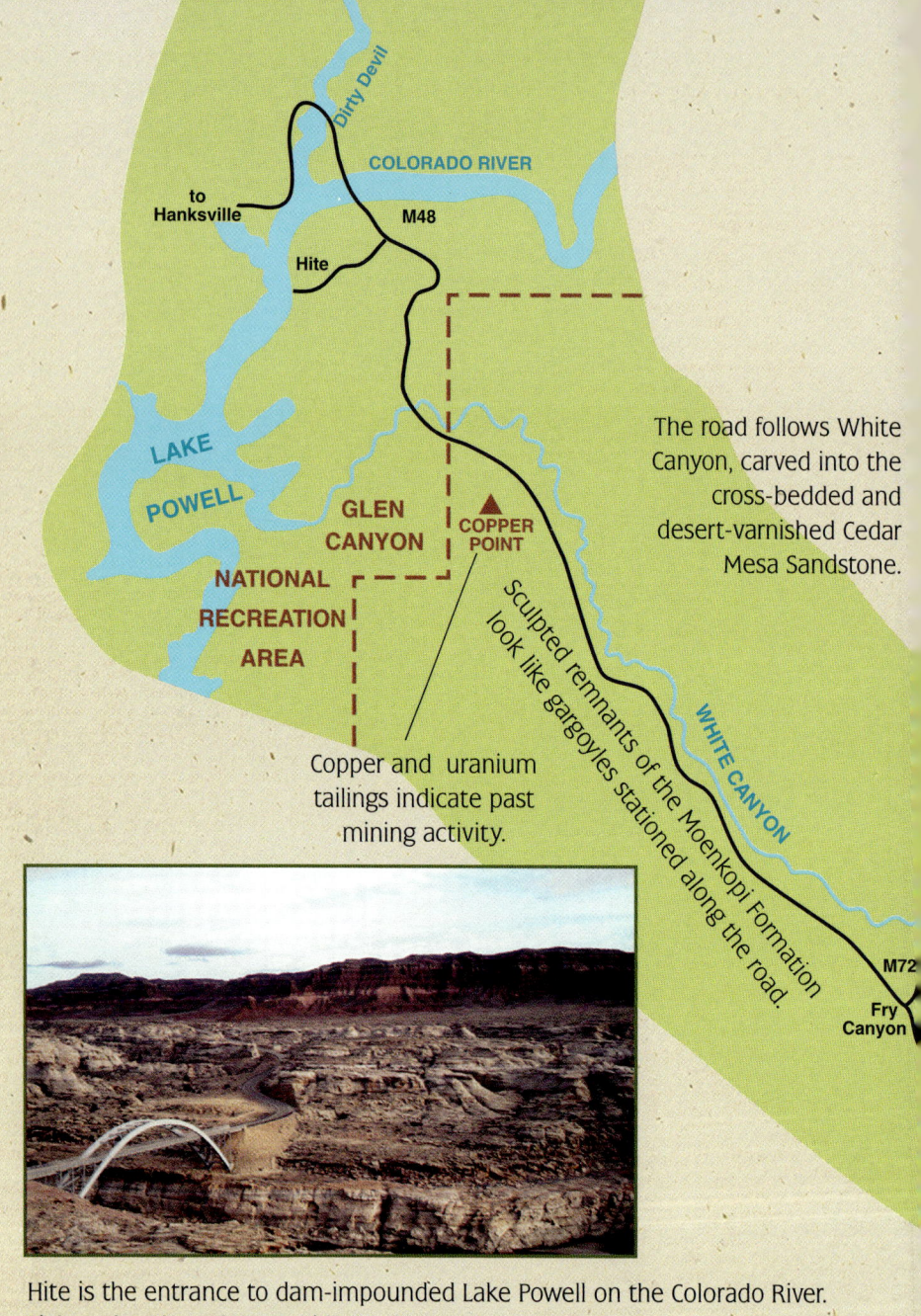

The road follows White Canyon, carved into the cross-bedded and desert-varnished Cedar Mesa Sandstone.

Sculpted remnants of the Moenkopi Formation look like gargoyles stationed along the road.

Copper and uranium tailings indicate past mining activity.

Hite is the entrance to dam-impounded Lake Powell on the Colorado River. Light-colored sandstone with the frozen-dune appearance is the Cedar Mesa Sandstone. Overlying red slopes and ledges (background) are the Organ Rock Formation, which contains a pink band of sandstone. The thin, white ledge capping the red slopes is the White Rim Sandstone, which noticeably thickens to the west. The Moenkopi Formation defines the skyline.

CANYON COUNTRY
A Land Of Enchantment

*T*his vast rock desert in southeastern Utah showcases deep, colorful canyons, majestic natural monuments pictured in western movies, a fantasy land of towering pinnacles and slot canyons, rock arches and bridges, and breathtaking vistas atop mesas. Much can be seen by car, and more of this endlessly intriguing country is accessible by foot, mountain bike, four-wheel-drive vehicle, or watercraft.

You do not have to be a geologist to enjoy or interpret the scenery, for the rocks easily divulge their stories of forming in seas or on land, in deserts or in deltas, in rivers or along beaches. What a fascinating heritage: dinosaurs once sauntered through lush lowlands leaving huge footprints in mud that has now turned to rock, vast sand dunes that drifted across the area are preserved as large sweeping lines in sandstone walls, and oceans that once covered this region left fossilized corals and seashells embedded in limestone.

This guide introduces you to the rocks' stories, to enhance your travels and your appreciation of the region. Along designated scenic byways, which are portrayed at the end of this pamphlet, look for the specific features described here and learn what their journey has been as you make yours. During your visit, tread lightly on this delicate land so that the pristine and unique landscape remains unchanged. Desert soils are extremely fragile; minimize your impact by staying on roads or marked trails.

Above photo and cover photo courtesy of the Grand County Travel Council.

Credits
Text by Sandra N. Eldredge. Photos by author unless otherwise credited. Graphics by Vicky Clarke. Special thanks to: the Salt Lake Convention & Visitors Bureau and the Utah Travel Regions Association for their support of this project; Canyonlands National Park and the Grand County Travel Council for photo releases; Newc and Sally Eldredge for field assistance; Miriam Bugden, Joan DeGiorgio, Hellmut Doelling, Kimm Harty, Carolyn Olsen, Michael Ross, and Christine Wilkerson for providing valuable reviews; and Patti MaGann who helped to get this project off the ground.

Time Passages

At many vistas in the Canyonlands region, the rocks display hundreds of millions of years of earth history. And yet, what you see are only small slices of earth's time, during which oceans repeatedly flooded and retreated, mountains rose and wore down, river systems appeared and disappeared, and sand dunes migrated across a Sahara-like desert.

Much of the landscape's uniqueness is due to underground salt movement. Salt collected in a restricted arm of an ocean 300 million years ago during the **Pennsylvanian** Period. At that time, the nearly flat landscape buckled into a mountain range (Uncompahgre uplift) and an adjacent basin (Paradox basin). The basin was intermittently connected to, and then isolated from, a nearby sea. During the isolation phases, salt water evaporated, eventually leaving vast salt deposits up to thousands of feet thick over most of what is now the Canyonlands region. These salts became the rock strata called the Paradox Formation.

The colorful rocks that are the essence of this canyon country were laid down as sediments, layer upon layer, over the salt during the next 150 million years in the Permian, Triassic, and Jurassic Periods. Early in **Permian** time, the sea started to recede to the west, leaving a coastal plain over the Canyonlands area. Rivers flowed from the Uncompahgre highlands over this coastal plain to the sea. Red sediments carried by the rivers accumulated in channels and flood plains. Concurrently, white beach sands deposited by the ocean blew eastward and formed inland sand dunes. Fluctuating sea levels and intermittent river activity across the region resulted in the interfingering of these red and white sediments, which hardened into rocks called the Cutler Group.

As the Uncompahgre mountains slowly eroded to low hills during the **Triassic**, the adjacent (southwestern) area evolved into tidal flats and a large river delta. Brown and red clays, silts, sands, and pebbles transported from the eroding mountains became the slope-forming rock strata known as the Moenkopi and Chinle Formations.

Drier climate during early **Jurassic** time transformed the region into a vast sandy desert; the wind-blown sands eventually hardened into rocks that now are the red and gold sandstone cliffs and sky-

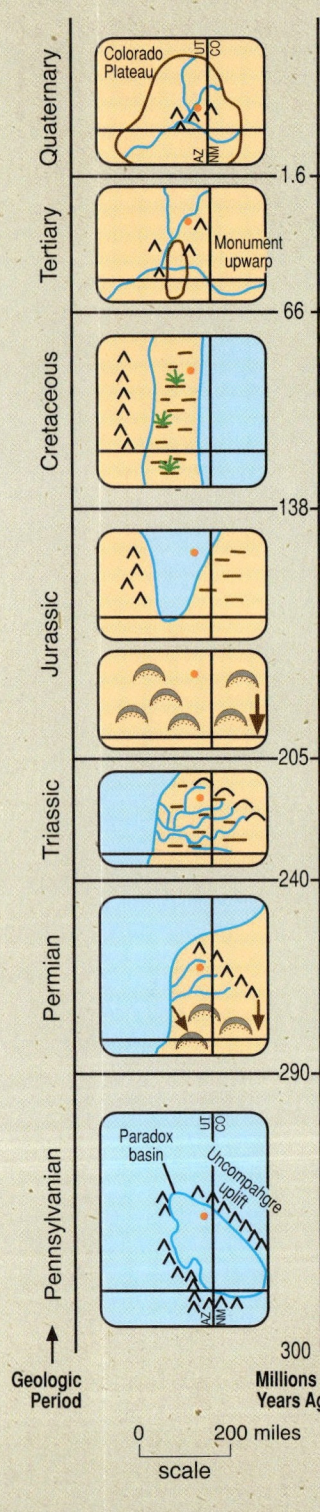

US-163 from the Arizona border to Bluff Scenic
45 miles, 1 1/2 hours

Travel across a 300-million-year-old ocean shelf complete with coral reefs and abundant marine life. This shelf formed the southeast margin of the Paradox basin. Much of the road traverses the Honaker Trail Formation, a gray limestone (weathered to a tan color under Mexican Hat Bridge) encapsulating fossils from the ancient ocean. Anticlines and synclines superimposed on the Monument upwarp were formed by compression of the earth's crust. This route passes through an enchanting landscape of monuments and spires and includes side trips to the Valley of the Gods and to impressive vistas via the Moki Dugway road. Half of the scenic byway is on the Navajo Indian Reservation.

A four-
the Mo
leads t
class
mear
cut d
through
age

Monument Valley Navajo Tribal Park is a landscape of fantastic mesas, buttes, and spires sculpted in the DeChelly Sandstone underlain by the red slopes of the Organ Rock Formation. The DeChelly originated as desert sand dunes from Permian time. The dunes were bordered on the northwest (approximate present location of the Colorado River) by the White Rim sand dunes and beaches along the fringes of a western ocean, and on the north (approximate present location of the San Juan River) by Cutler Group rivers.

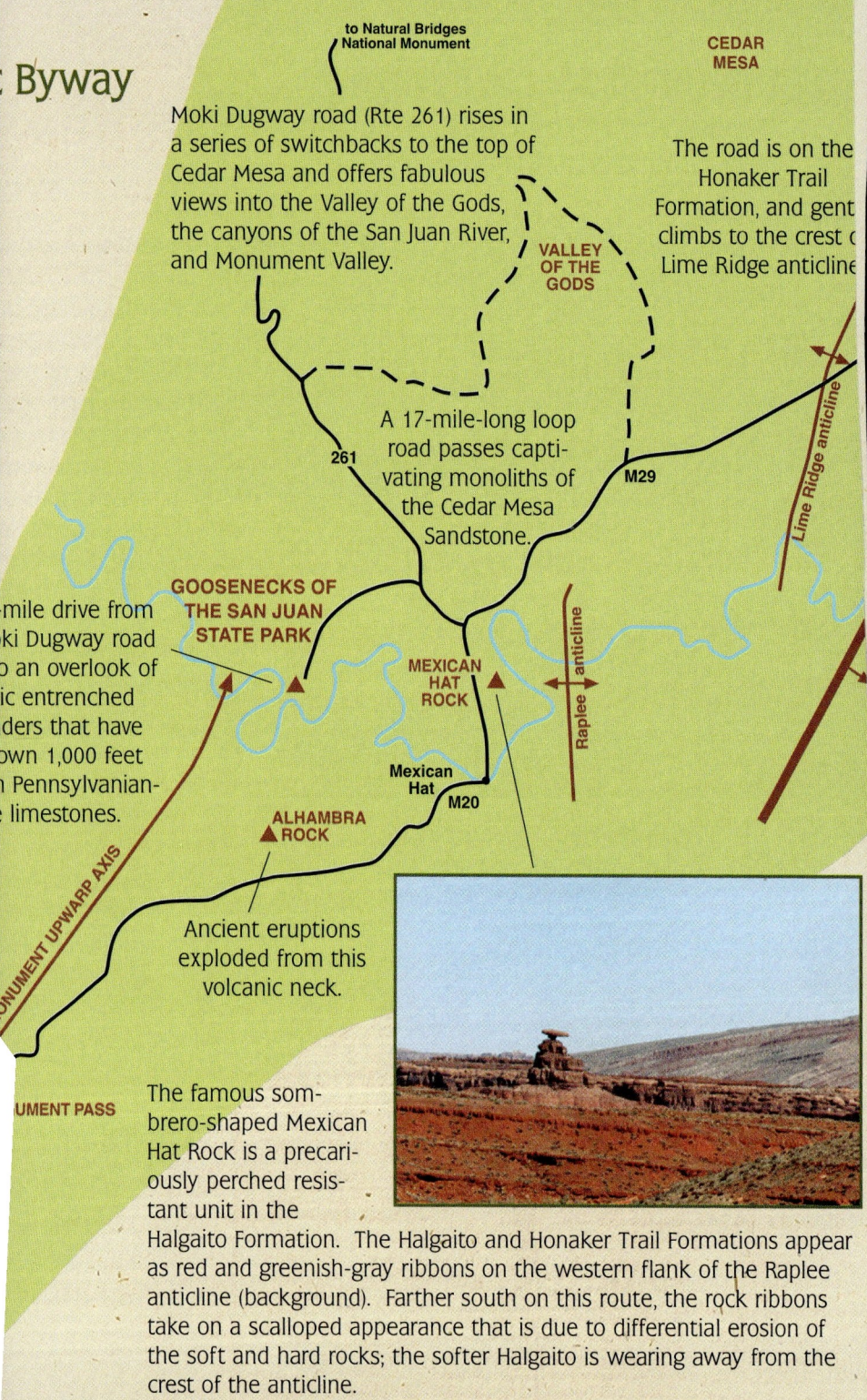

Byway

to Natural Bridges National Monument

CEDAR MESA

Moki Dugway road (Rte 261) rises in a series of switchbacks to the top of Cedar Mesa and offers fabulous views into the Valley of the Gods, the canyons of the San Juan River, and Monument Valley.

The road is on the Honaker Trail Formation, and gent climbs to the crest of Lime Ridge anticline

VALLEY OF THE GODS

A 17-mile-long loop road passes captivating monoliths of the Cedar Mesa Sandstone.

261

M29

GOOSENECKS OF THE SAN JUAN STATE PARK

-mile drive from oki Dugway road o an overlook of ic entrenched ders that have own 1,000 feet Pennsylvanian- limestones.

MEXICAN HAT ROCK

Raplee anticline

Lime Ridge anticline

Mexican Hat

M20

ALHAMBRA ▲ ROCK

Ancient eruptions exploded from this volcanic neck.

MONUMENT UPWARP AXIS

UMENT PASS

The famous sombrero-shaped Mexican Hat Rock is a precariously perched resistant unit in the Halgaito Formation. The Halgaito and Honaker Trail Formations appear as red and greenish-gray ribbons on the western flank of the Raplee anticline (background). Farther south on this route, the rock ribbons take on a scalloped appearance that is due to differential erosion of the soft and hard rocks; the softer Halgaito is wearing away from the crest of the anticline.

to
Blanding
191
NAVAJO
TWINS
M45
Bluff
(see Rte 95
for description)
163
COMB RIDGE
San Juan River
191

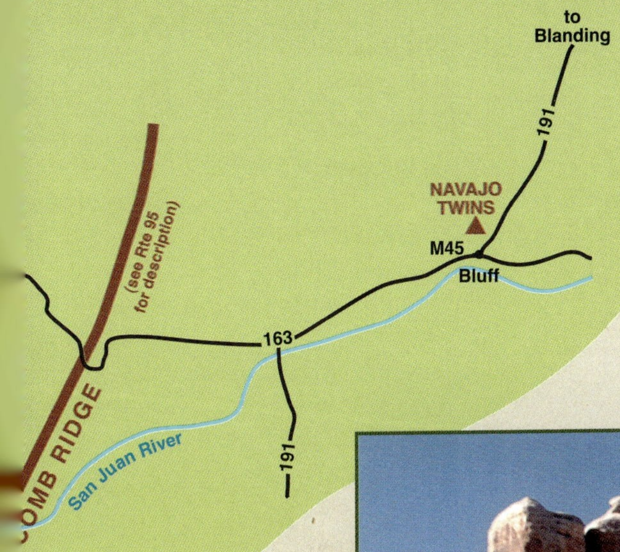

The local, wind-deposited Bluff Sandstone (a member of the Jurassic-age Morrison Formation) is named after this town.

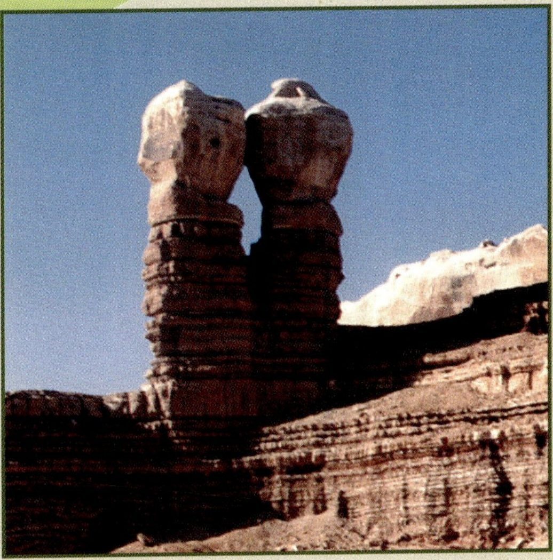

The Navajo Twins are caps of Bluff Sandstone.

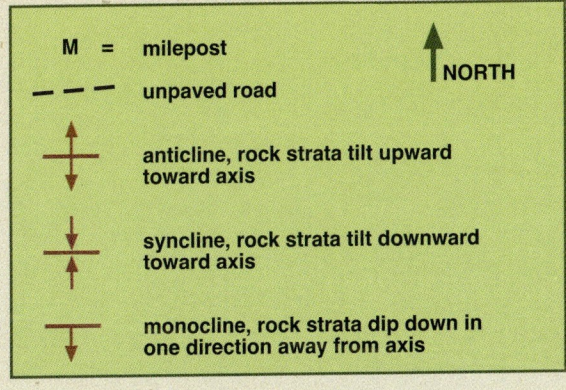

M = milepost
unpaved road
NORTH
anticline, rock strata tilt upward toward axis
syncline, rock strata tilt downward toward axis
monocline, rock strata dip down in one direction away from axis

Information Agencies

Salt Lake Convention & Visitors Bureau
180 South West Temple
Salt Lake City UT 84101 (801-521-2822)

Utah Travel Council
Council Hall, Capitol Hill
Salt Lake City UT 84114 (801-538-1030)

Moab Information Center
Center St. and Main St., Moab UT 84532
(Staffed by Utah Travel Council, Canyonlands Natural History Association (801-259-6003), Bureau of Land Management (801-259-8193), National Park Service, and U.S. Forest Service)

Arches National Park
P.O. Box 907, Moab UT 84532 (801-259-8161)

Canyonlands National Park
125 West 200 South, Moab UT 84532 (801-259-7164)

Grand County Travel Council
210 North 100 West, Moab UT 84532 (1-800-635-6622)

Dead Horse Point State Park
P.O. Box 609, Moab UT 84532 (801-259-6511)

San Juan County Travel Council
117 South Main St.
Monticello, UT 84535 (801-587-3235)

Edge of the Cedars State Park
P.O. Box 788
Blanding, UT 84511 (801-678-2238)

Monument Valley Navajo Tribal Park
P.O. Box 360289 (801-727-3287/3353)
Monument Valley, UT 84536

Natural Bridges National Monument
Box 1, Lake Powell, UT 84533 (801-259-5174)

U.S. Geological Survey
Earth Science Information Center
2222 West 2300 South
Salt Lake City, UT 84119 (801-975-37420)

Utah Geological Survey
2363 South Foothill Drive
Salt Lake City, UT 84109 (801-467-0401)

Contact the Utah Geological Survey for other geologic publications, including: Geology and Grand County, Geologic Resources of San Juan County, and the Geologic Map of Arches National Park.

FEATURES ON CANYONLANDS TRAVEL REGION MAP
(back cover)

Book Cliffs-described in Time Passages
Uncompahgre uplift-described in Time Passages
Upheaveal Dome, a mysterious geological feature, one theory is that it is an astrobleme, or giant meteor crater
The Needles-described in Salt Movement and Rte 211
The Grabens-described in Salt Movement
Monument upwarp-described in Time Passages and Rte 95
Paradox basin-described in Time Passages
Newspaper Rock State Park, Edge of the Cedars State Park, and **Hovenweep National Monument** are situated around Indian ruins and petroglyphs.

Island in the Sky District of Canyonlands National Park and **Dead Horse Point State Park** (inset) offer inspiring views of nature's red artwork and the erosive power of the Colorado River. The White Rim Sandstone forms the terrace 1,000 feet below the Island in the Sky, upon which the 100-mile-long **White Rim Trail** can be traversed by mountain bike or jeep.

CANYONLANDS TRAVEL REGION

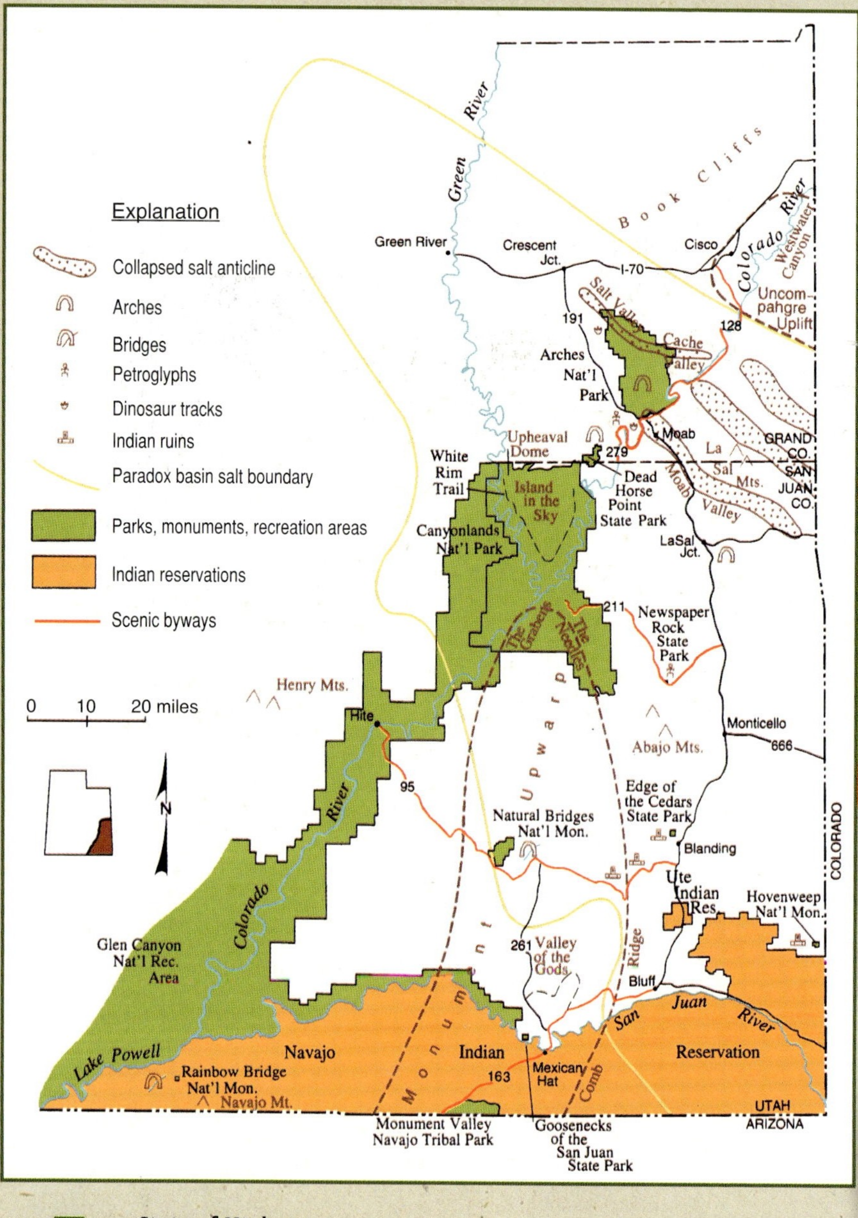

State of Utah
Department of Natural Resources
Utah Geological Survey
Public Information Series 34
1996 ISBN 1-55791-373-0

ISBN 1-55791-373-0

Canyon Country